Reach

Building Communities and Networks
for Professional Development

Jeff Utecht

Copyright © 2010 by Jeff Utecht
Cover design by Jeff Utecht
Cover Image by Flickr.com user Woodleywonderworks

Book design by Jeff Utecht
This work is licensed under the Creative Commons Attribution-Noncommercial-Share Alike 3.0 United States License.
To view a copy of this license, visit http://creativecommons.org/licenses/by-nc-sa/3.0/us/ or send a letter to Creative Commons, 171 Second Street, Suite 300, San Francisco, California, 94105, USA.

Jeff Utecht
Visit my website at www.jeffutecht.com

Printed in the United States of America
First Printing: June 2010
ISBN-13 978-0-615-38118-3

Contents

Preface   **I**

Acknowledgments   **III**

Introduction   **1**

    What is Web 2.0   1

    The Rapid Growth of Technology   2

    Globalization Education   3

    Education Today   5

1. Communities and Networks   **7**

    Networks   8

    Your Personal Learning Network   10

    Networks without Communities   11

2. Building Your Network   **17**

    RSS   18

    Becoming Part of a Community   26

    Growing Your Professional Network   31

    Making Your Network Work for You   36

3. Tools to Build a Community	**41**
Blogs	41
Wiki	50
Ning.com	51
4. Network Building Tools	**54**
Twitter	54
Facebook	61
Skype	66
Google Buzz	71
5. Connecting It All Together	**73**
Living in the Cloud	73
6. Making it Personal	**81**
How Far is Your Reach?	85
References and Links	**86**
Reach Website	**93**
About the Author	**94**

Preface

It was early August 2005, as I sat in the computer lab waiting for the next class of Kindergarteners to make their way down the hall for our weekly computer lesson. I browsed the Internet looking for resources and project ideas that I could use with the fifth graders I would have the following day. As I sifted through the countless resources on the web, I came across a blog. I had heard of blogs of course, and had even signed up for a free one on blogger.com a few months before. But up until this point, I had never stopped to actually read an educational blog. So I did just that, I began to read. And as I read, I found myself agreeing with many of the author's points and ideas. As I read further, I discovered that the author of this blog was writing about something he had read on another blog. Clicking on the link, I found myself on my second educational blog. Reading more, agreeing more, and learning more, I soon found myself spending much of my prep time reading blogs. A couple of weeks later, I actually worked up enough nerve to leave my first comment on someone's blog. A month later, I signed up for an RSS reader and a delicious.com account.

    By the middle of September, my head was spinning with the new ideas, tools and experiments I was reading about and trying with my own students. I began to play around with the idea of creating my own blog. I knew I wanted to share my experiences, my excitement about what I was learning with others. I felt a need to give back to the community of educators I was now learning from on a daily basis. There was one massive glaring problem, however, I am not a writer. I have never enjoyed writing, or reading for that matter.

I have a learning disability and struggled with reading, writing and school as a whole, most of my life. But still, I found myself constantly thinking about creating my own blog. Who would read my thoughts? Wouldn't people make fun of all my spelling mistakes? What if someone doesn't agree with me? For weeks these questions ran through my head until one day, I read a blog post on how to install your own blogging software. I downloaded the program, followed the steps, and 30 minutes later I had installed my first blog. With this hurdle done, now I had to figure out what to name it. Looking around my room, I spotted my baseball bat in the corner. This was the bat that my best friend had given me my first year of teaching. I kept it in my classroom and would hold it in my hands when I was doing my deepest thinking and brainstorming with my students. This bat was with me still, in my third country and sixth teaching space. I simply typed "The Thinking Stick" into the title box, clicked save and my life was irreversibly changed.

    The educational community that I would become a part of that day changed the way I viewed students, schools, and education. I had become a node of knowledge in a network of learners. With one simple click of the button, I had created my little piece of the web that would allow me to communicate and collaborate with others who were all passionate about the same thing: learning. I had entered the world we now know as Web 2.0.

Acknowledgments

The author would like to thank the following individuals for their assistance:

Dana Watts
High School English Teacher
International School Bangkok
Bangkok, Thailand
http://www.teachwatts.com/

Woodleywonderworks
Front Cover Picture
http://www.flickr.com/photos/wwworks/3195267875/

To all the readers of The Thinking Stick Blog, thank you for shaping this book and showing me the power of online communities and networks.

To my wife, thank you for your patience and constant support. We both know this book would not have happened without your love, your commitment, and your endless belief in me.

*To My Kleine Taube
Because Spring Comes on Forever*

Introduction

What is Web 2.0

Web 2.0 is a term that we hear a lot these days. But it is also a term that we find hard to explain just what it means. The discussion around exactly what the term means is still being debated. The Wikipedia article for the term Web 2.0 is an interesting place to start http://en.wikipedia.org/wiki/Web_2.0.

One aspect of Web 2.0 that everyone has agreed upon though, is that it allows users to engage with information and people. When explaining what I believe Web 2.0 means, I like to use the analogy of a newspaper.

Newspapers show up on our doorsteps or in the paper box every day. We read the articles within them, we think about what is being written, but we never have an opportunity to discuss our thoughts with the journalists who write these articles. The information flows one-way, from the journalist to us. There is no way for the reader to discuss the article with other readers; there is no way to ask clarifying questions to the journalist. The newspaper is Web 1.0, defined as: Information that flows from producers to consumers of knowledge.

Web 2.0 allows conversations. It allows that newspaper article to be talked about and to be discussed. What if you could read a newspaper article and leave a comment back to the journalist that wrote it? What if you could talk about the article with other people around the world that read the same article as you? What if you could edit the article? Or even create an article yourself on something you were interested in? No longer would you have to go to the local coffee shop in hopes of finding someone who read the same article to have a discussion. You could have all those conversations and more with the click of a button.

Web 2.0 allows anyone to create content. No longer are there consumers and producers. We have all become prosumers (Wikipedia, 2009): Creating knowledge as we consume.

In 1985, four years before the first web browser as we know them today was invented, the creators of the web understood the true power of the Internet. They knew it was not in the content that would one day be stored there, but in the communities that it would create. Steve Case, one of the engineers considered to be one of the founding fathers of the Internet, stated in an interview, "We thought communities trumped content" (Mayo & Newcomb, 2008). Before Web 2.0 was even a thought in our minds, the founding fathers of the Internet believed that it would be communities of prosumers that would power the Internet.

The Rapid Growth of Technology

While driving through the deserts of Saudi Arabia in 2003, I passed a man on a camel talking on his cell phone. It was at that moment I realized that communication technology was increasing at such a rapid pace that whole countries were leapfrogging technology inventions. Very few people in Saudi Arabia had telephones in their homes, yet in the middle of the desert here was a man who lived in a tent with his family and was connected to others.

Today there are over 3 billion cell phones in the world. That is half the world's population and the number of cell phones continues to grow at a rapid rate. (Ahonen, 2008)

The One Laptop Per Child project (OLPC) from the Massachusetts Institute of Technology decided to produce the low cost child friendly laptop with wireless and not wired capabilities. What Nicholas Negroponte, the founder of the project, realized was that the growth of wireless communication around the world continues to out pace wired systems. In many cases, as in the rural areas in Thailand, small villages can receive a wireless Internet signal before they can a wired one (see laptop.org).

Globalization Education

Being an International Educator, I have the opportunity to not only see globalization first hand, but also be a part of it. The hundreds of international schools around the world serve two primary purposes:

1. To educate the expatriate child who's family has moved as part of the parents' global jobs or careers.
2. To educate those students who plan to attend the top colleges and universities around the globe.

I've worked at schools with children whose parents were part of the oil business, car manufacturers, computer manufacturers, and many other global corporations. Over the past 10 years, international schools have seen record numbers of new students. At the school I worked at in Shanghai, for example, the school went from serving 800 expatriate students (students with passports other than Chinese) to over 2000 in just a few years time.

Although many people think of globalization as the outsourcing of jobs, I have another view of globalization, a view of centralized placement.

While living in Shanghai, I met an American who worked for IBM. He had a BA in Art Foundations. As we talked, I learned that he was based in Shanghai and his job was to communicate with the software developers in India, the hardware developers in China, and the sales force in the United States. Together, they created bundles of software to be sold on computers to consumers. He talked about the ability to communicate with all parties using Skype for free and that although IBM had provided him with a nice office space, he only visited the office to pick up his mail. When I asked him how a person with a BA in Art ends up with a job in the computer/communication business, he talked about his interview with IBM. What IBM was looking for was somebody who could learn, unlearn and relearn quickly. The degree was less important than the skills he had as a communicator and learner. Although everyone he communicated with was in a different part of the world, he lived in Shanghai as it was central to his communication needs with all parties.

The distribution of jobs such as this one has been made easier due to technology. Skype.com, the free Internet telephone service, has made staying in touch with friends and family easier and cheaper. A computer to computer phone call from Bangkok, Thailand to my parents' home in Spokane, Washington is now virtually free (cost of Internet connection not included). A report released in January 2010 showed that Skype now makes up 12% of all international calls, more than any other single telephone company (Telegeography, 2010).

The true power of the Internet can be found in communities that form just in time around any given topic. The recent outbreak of the H1N1 virus is a perfect example of how the Internet is fostering communication and the creation of communities to keep track of research, locate outbreaks, and poll resources together to find a vaccine. Researchers globally now have access to each other's data. Consumers can keep track of outbreaks on a Google Map created by a community of concerned global citizens (http://flutracker.rhizalabs.com/). This is true globalization in action for the betterment of humanity.

Education Today

As our world continues to shrink due to globalization and as online communities continue to take hold, the learning potential that the web offers continues to grow. There is more and more content being put on the Internet each day. Content that we, as educators, need to teach our students how to consume, remix and produce. Whether we like it or not, or agree with it or not, the Internet is becoming the place we turn to for information. Yet, very few of our classrooms are teaching students how to consume and produce information in this new medium. Are we teaching our students to become prosumers?

If we are to teach our students to become prosumers of information in today's connected digital world, then we need to understand and become prosumers ourselves. By reaching out and joining online communities, create learning networks, and grow those networks to be powerful professional learning environments, educators can take advantage of the wealth of knowledge on the web. They can use this new knowledge for their own professional growth and pass the knowledge and power of the network on to their students.

1 Communities and Networks

Communities are nothing new to humans. We have been creating them since the beginning of time. Answers.com gives one definition of a community as follows:

A group of people having common interests

The neighborhood in which you live is one example of a community. You have a common interest of living in the same area as your neighbors. You might not know all of them, or even like all of them, but you belong to a group of people with a common interest, in this case location.

The same applies to communities on the Internet. Using one of the largest online communities Facebook.com as an example, we can see how online communities are developed. There are over 400 million users on Facebook; they are a community of users who have joined that site, created profiles, uploaded pictures, and communicate regularly with people they know. As a user of Facebook you become a part of that community. You have a common interest in connecting with other people you know, or connecting with others that share a common personal or professional interest. Once you create an account you are welcomed into the Facebook community. What you do and how involved you are once you get there is up to you.

Networks

Once you join a community you are able to create a network within it that will serve your purpose. Answers.com gives a definition of networks as being:

An extended group of people with similar interests or concerns who interact and remain in informal contact for mutual assistance or support.

This is the type of network the Internet allows. Once you join a community you then create your network within it by reaching out and connecting with others. Facebook works on this principal. Joining the community of Facebook gives you nothing. Creating a network of friends and family is the power that lies within the Facebook community. You will never know the millions of members in the community called Facebook. You won't interact with most of them. Much like the neighborhood you live in, or the grocery store you shop at, they are there going about their daily lives in your community and you within theirs. But within that community you have the potential to find people you care about, people you can learn from, and friends you enjoy being around. Facebook is the community that allows the connections to happen between community members to form networks.

This Facebook group for the newly released product Google Wave (Figure 1.1) is one way networks are formed on Facebook. There are 21,139 members in this network. They have decided to join this network because they have an interest in Google Wave. This represents a network within a community. Within the community of Facebook there is a niche network of members who have reached out to each other to interact and share information with other members who are interested in Google Wave. Without being a member of the Facebook community you can not be a member of the Google Wave Group network.

Figure 1.1

Twitter.com is another example of the community/network relationship. Creating an account on Twitter along with millions of other users does nothing more than give you the potential to interact with some of its members. The network of people you decide to connect with inside that community will determine how meaningful that community is to you. The more meaningful the connections you make within a community, the more powerful that community and the network you create become to you.

This is what "social-networking" means on the web. It is the ability of users to create a social network that is meaningful to them. My social-network is personal to me. The people I "follow" or "friend" creates my unique network. I've created a social network of like-minded people who I consider friends, colleagues, and partners in learning and can reach out to them for advice, lesson plans, or resources.

Your Personal Learning Network

The name given to the networks that you create that revolve around personal and professional development are called Personal Learning Networks or PLNs. The network of people you decide to become connected to is personal to you. There is no other member of Twitter that has the same network as I do. That network is personal to me. I have decided to connect with specific people within that community I feel I can learn from in a professional manner. They are my "Learning Network" in that I learn and gather information from them on a daily basis. From links to articles, to ideas on lesson plans, my Personal Learning Network is a real time professional development network of educators that I rely on to help me do my job as an educator.

Networks without Communities

Not all social-networks require you to join a community first. There are tools on the Internet that allow you to create a network of people and information that is personal to you. Using Really Simple Syndication or RSS you can create a personal network without having to join a specific community. RSS is a simple Internet format that allows a user to connect a website to what is called an RSS Reader. This reader then reads the code and displays the information within the RSS Reader. Commonly used RSS Readers are Google Reader, Netvibes.com and Pageflakes.com. All three of these web sites do the same thing in that they allow a user to pull information from a number of sites using RSS to display it in one location. Simply sign into your RSS Reader and have access to multiple sites. Creating a Personal Learning Network this way does not mean you first join a community; instead you create your learning network from sources around the web. It could be your favorite blogs, newspapers, magazines, or any other site that offers its consumers the ability to use RSS. What RSS allows a person to do is change the flow of information on the web. Instead of constantly going to your favorite website to see what's new, anytime something is new on your favorite websites, it comes to you through your RSS reader.

RSS just might be the most powerful technology in the Web 2.0 revolution. The idea of creating a network of information from multiple sources and have that information find you, instead of you finding it, changes the professional development landscape. You control the content that comes to you. By creating a one-time connection between a web site and your RSS reader you now control the information that finds you.

Creating a learning network using RSS allows you to create your personal network of information without having to join a community of people. This is the difference between joining a community and creating a network, or creating a network without joining a community. Twitter and Facebook are about connecting to people and the content that they create. An RSS reader and a network without a community is about creating a network of content on a given topic or idea. Using both within your greater Personal Learning Network creates a powerful platform for learning and growing professionally.

Building your RSS Reader

I recommend using a web-based RSS reader that will allow you to connect to your content from any device that has an Internet connection. There are many options out there and finding one that fits your reading style and needs is what is most important. You might have to try a couple different RSS readers to find the one that fits you.

Google Reader (www.google.com/reader)

Google Reader is a popular RSS reader that has many great features. If you already have a Gmail or Google Account getting started with Google Reader is as easy as going to google.com/reader and logging in with you google account.

Positive Features:
- Uses folders to organize RSS feeds
- Integrates with iGoogle
- Quick loading
- Use the power that is Google
- Allows you to connect with friends and share your favorite reading with them and theirs with you.
- Stories appear in order of when they were published or by individual site.

Negative Features:

- Supported by Google Ads

Videos to get started:

- http://tinyurl.com/googlereaderintro
- http://tinyurl.com/googlereadersetup

Pageflakes (www.pagefllakes.com)

 It differs from Google Reader in that it has "Flakes" which are items you can add to your page other than just RSS feeds. Pageflakes has a great little Podcast player that you can add and use to listen to podcasts. You can add the weather, news, a search box and other widgets and customize your start page to be the information center you want it to be.

Positive Features:
- Allows you to add more than just RSS Feeds.
- Create Tabs (rather than folders) to keep your information organized.
- Allows you to connect to Facebook, Twitter and other networks bringing all your information to one place.
- Make the page public to share with your class.

Negative Features:
- Ad supported with "Sponsored Ads"
- Can be slow to load depending on what widgets you have on your page.

Videos to get started:
- http://tinyurl.com/pageflakesintro
- http://tinyurl.com/pageflakesshare

Netvibes (www.netvibes.com)

My personal favorite Netvibes was the first to use widgets to really make a customized start page on the web that allowed a user to import your own RSS feeds. They continue to add widgets that allow you to add functionality to your page. Netvibes uses tabs to allow you to organize your content. You can also share a tab with others. You can set up students with Netvibes pages and then share a tab that has all the RSS feeds you want them to read. Or create a public page (www.netvibes.com/jutecht) and share with others what you're reading and other information. I use Netvibes with students because they are a customizable generation and Netvibes lets you customize everything.

Positive Features:
- No ads
- Allows you to add more than just RSS Feeds.
- Create Tabs (rather than folders) to keep your information organized.
- Allows you to connect to Facebook, Twitter and other networks bringing all your information to one place.
- Make the page public to share with your class.
- Share a specific tab with other Netvibe users
- Easy to see what you have read
- No need to leave Netvibes to view the site you are reading

Negative Features:
- Can load slow if you have a lot of content

Videos to get started:
- http://tinyurl.com/netvibesintro
- http://tinyurl.com/netvibessetup

Spend some time watching these videos and finding the RSS reader that fits your way of gathering and reading information. Once you have your RSS reader created make it your browser's homepage and get in the habit of just having it open when you're doing anything on the Internet.

Once you have created a stream of information that finds you rather than you spending your time finding it, you'll be amazed how much less time you spend searching for information. This provides you with more time to read and reflect about the information that matters to you most.

A RSS reader is never done being built. Continue to add and delete feeds based on your needs and voices on the Internet that speak to you.

2 Building Your Network

Now that we have discussed the differences between networks and communities, I think it is worth discussing what you will have to do once your network is set up and the connections are made. Setting up your network is the easy part. Making it work for you takes effort and time. If a personal network is going to fail, it is in part due to the time and effort that one allows himself/herself to spend within the network reaching out and learning from others. Much like participation within a classroom, the more you participate the more you get out of the class. You need to be a willing participant within the larger conversation that takes place within your newly joined community. Much like a classroom discussion, you must be an active member to get the most out of the content being shared.

Once you join a professional community you become a node of information and ideas within that community for others to connect with. Although you set up online networks to allow information to flow to you, being an active part of the community is important and helps to establish those connections for your learning network. Being what is commonly referred to as a 'lurker' within online communities is a great way to start. Someone who is just there watching and learning, but does not give back or actively shares within the community itself is all fine, but the real learning and relationships and network building comes when you become an active part of the community. If everyone were a lurker, no information would be shared. Sharing and being active within the community is the first step.

The more active you are within a community the more visible you become to other members. The more visible you become, the more potential connections are created.

Activity = Visibility = Connection opportunity

It is those connections that are created by you reaching out and connecting and communicating with others that lead to the deeper learning that happens within a personal learning network.

As mentioned earlier, I believe that RSS (Really Simple Syndication) might just be the best tool that has emerged out of the Web 2.0 revolution. I recommend starting to build your network around an RSS reader. Using RSS, as discussed in Chapter One, also allows you to create your personal learning network without joining a community. No need to worry about your avatar (small icon picture that comes with a community profile) or how and with whom you should connect. Starting with an RSS reader and then moving into a community of learners is my recommended process.

RSS

I will do my best to explain in more detail how RSS works here, but I understand that some of you might learn better by watching a video or reading a paper by someone else. So I have gathered videos and resources to help you understand RSS a bit better at this book's website jeffutecht.com/reach.

Setting up and getting started with an RSS reader is as easy as Copy and Paste. The hard part is knowing what to copy and where to paste. RSS is basically a connection. Let's walk through a scenario of how to create this connection between an RSS enabled website and your RSS Reader.

Jane is looking for resources on Global Warming. She's getting ready to start a new unit on Global Warming in her science class, and wants to find some sites that she can use to start conversations in her classroom. She wants current information on global warming, but doesn't want to spend hours of her precious planning time looking for current news. She decides to start a new folder in her Google Reader and names it Global Warming. She then takes one planning period looking for sites that will give her news about Global Warming and have an RSS feed that she can add to her reader. She first heads to Google News http://news.google.com and does a search for "Global Warming". Jane knows that Google News pulls information from over 4500 newspapers all around the world. By doing a search for Global Warming at Google News she is getting the latest news articles about Global Warming from places such as China, Australia, Brazil, and America. She can use these different perspectives to spark conversations in her classroom.

After doing the search, Jane scrolls to the bottom of the results page and finds a link that reads RSS. She clicks on the link, which loads a web language looking page. She's

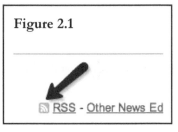

Figure 2.1

not worried about the page as she knows that the information on the page is just information for her Google Reader to read. What she really wants is the URL or Address. She highlights and copies the URL of this page, she then opens her Google Reader, clicks on the blue and gray "Add a Subscription" button and pastes the URL from her Google News search into the space provided. She then clicks OK and in a second the latest news is imported into her reader. Now as new articles are written that talk about Global Warming, her Google Reader will be automatically updated with the latest articles. She has her Google Reader set up to import new data every 5 minutes so she is always getting the latest updates from Google News.

Next she heads over to delicious.com, a social bookmarking website where people save their bookmarks, tag them, and make them available to users on the web to search and use. Jane has used delicious.com before and knows exactly what she is looking for. She types http://www.delicious.com/tag/globalwarming into her web browser to find articles/websites tagged with the word globalwarming.

She is given a list of websites that the 3 million users of delicious have decided to tag (or categorize) with the word "globalwarming". She then scrolls to the bottom and finds the little orange RSS button.

She clicks on it, copies the address, loads her Google Reader, clicks on the blue and gray "Add a Subscription" button and pastes that address into the space provided and adds the feed. Now whenever a delicious user finds a new site and decides to tag it with the word "globalwarming" Jane's Google Reader will tell her that there is a new article/website that has been added by one or more delicious.com users. Jane will have no need to return to Google News or to Delicious.com again in her quest for resources on Global Warming, the information from those sites will now come to her in one location with constant updates. Jane's Google Reader is her one stop shop to what interests her on the web.

RSS is nothing more than a connection between your RSS Reader and a website of information. It's becoming very popular on the web with websites that change information frequently such as Blogs, Newspapers, Magazines, and updated websites. Many social networks use RSS as well. Ning.com, Twitter.com, Delicious.com, and Diigo.com just to name a few allow anyone to make a connection between the content found on their sites and your personal RSS Reader. An RSS Reader acts like a gateway to the web. After the connection to the website has been made, your RSS Reader 'fetches' the information from the website and delivers it to you. No longer are you running around the web searching for new information or wondering if a site has been updated. Your Reader does the work for you; all that you have to do is read the content.

Building your RSS Reader

I recommend using a web-based RSS reader that will allow you to connect to your content from any device that has an Internet connection. There are many options out there and finding one that fits your reading style and needs is what is most important. You might have to try a couple different RSS readers to find the one that fits you.

Google Reader (www.google.com/reader)

Google Reader is a popular RSS reader that has many great features. If you already have a Gmail or Google Account getting started with Google Reader is as easy as going to google.com/reader and logging in with you google account.

Positive Features:
- Uses folders to organize RSS feeds
- Integrates with iGoogle
- Quick loading
- Use the power that is Google
- Allows you to connect with friends and share your favorite reading with them and theirs with you.
- Stories appear in order of when they were published or by individual site.

Negative Features:

- Supported by Google Ads

Videos to get started:

- http://tinyurl.com/googlereaderintro
- http://tinyurl.com/googlereadersetup

Pageflakes (www.pagefllakes.com)

It differs from Google Reader in that it has "Flakes" which are items you can add to your page other than just RSS feeds. Pageflakes has a great little Podcast player that you can add and use to listen to podcasts. You can add the weather, news, a search box and other widgets and customize your start page to be the information center you want it to be.

Positive Features:
- Allows you to add more than just RSS Feeds.
- Create Tabs (rather than folders) to keep your information organized.
- Allows you to connect to Facebook, Twitter and other networks bringing all your information to one place.
- Make the page public to share with your class.

Negative Features:
- Ad supported with "Sponsored Ads"
- Can be slow to load depending on what widgets you have on your page.

Videos to get started:
- http://tinyurl.com/pageflakesintro
- http://tinyurl.com/pageflakesshare

Netvibes (www.netvibes.com)

My personal favorite Netvibes was the first to use widgets to really make a customized start page on the web that allowed a user to import your own RSS feeds. They continue to add widgets that allow you to add functionality to your page. Netvibes uses tabs to allow you to organize your content. You can also share a tab with others. You can set up students with Netvibes pages and then share a tab that has all the RSS feeds you want them to read. Or create a public page (www.netvibes.com/jutecht) and share with others what you're reading and other information. I use Netvibes with students because they are a customizable generation and Netvibes lets you customize everything.

Positive Features:
- No ads
- Allows you to add more than just RSS Feeds.

- Create Tabs (rather than folders) to keep your information organized.
- Allows you to connect to Facebook, Twitter and other networks bringing all your information to one place.
- Make the page public to share with your class.
- Share a specific tab with other Netvibe users
- Easy to see what you have read
- No need to leave Netvibes to view the site you are reading

Negative Features:
- Can load slow if you have a lot of content

Videos to get started:
- http://tinyurl.com/netvibesintro
- http://tinyurl.com/netvibessetup

Spend some time watching these videos and finding the RSS reader that fits your way of gathering and reading information. Once you have your RSS reader created make it your browser's homepage and get in the habit of just having it open when you're doing anything on the Internet.

Once you have created a stream of information that finds you rather than you spending your time finding it, you'll be amazed how much less time you spend searching for information. This provides you with more time to read and reflect about the information that matters to you most.

A RSS reader is never done being built. Continue to add and delete feeds based on your needs and voices on the Internet that speak to you.

Become Part of a Community

Communities are formed through communication. What makes a community of people worth joining is the communication between all people within that particular community. Online, each person is a node of information, that is, a point of contact. If that point of contact, that person, is helpful and shares good ideas, they become a more important node. If a person, or information node, does not share or communicate they become less important to the community until the point that the community moves on without them. In order to be a node within a community you must be an active contributor. Of course being an active member means dedicating time to the community and reaching out to create your own network within it. If you want the network to work for you when you need it, you need to be there for others when they need you.

Being active in a community is where many educators start to feel overwhelmed. Educators have enough to do managing their own classrooms on a daily basis without having to be an active member to one or more online professional communities. This is where many educators get stuck and give up. They create a network within a community, use it for a day or two and then never return. Much like muscles that are not exercised, those connections become weak until they whither away completely. Educators might then come back a month later after hearing from others the power of a personal learning network, only to find that they have nothing.

From: http://www.thethinkingstick.com/digital-literacy-vs-networked-literacy

Published On: August 7, 2009 edited for this book

I woke up this morning to find the following Tweet from Jeremy Brueck (`http://twitter.com/brueckj23`):

> Wondering wht the difference between 'Digital Literacy' & 'Networked Literacy' is/might B? http://tr.im/vKOk @jutecht #canuhelpmeunderstand?

This line between digital literacy and networked literacy is a fine one...but one I think is worth exploring.

I first started thinking about the distinction between digital literacy and networked literacy after reading the Writing in the 21st Century (`http://www.ncte.org/press/21stcentwriting`) document produced by the National Council of Teachers of English (`http://www.ncte.org`) and Kathleen Yancey.

In the document Yancey states:

First, we have moved beyond a pyramid-like, sequential model of literacy development in which print literacy comes first and digital literacy comes second and networked literacy practices, if they come at all, come third and last.

Based on this reading and more specifically this paragraph, I created this diagram:

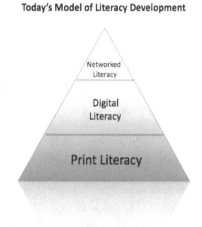

I wrote on this topic in an article titled The Age of Composition (http://www.thethinkingstick.com/the-age-of-composition). Yet in that article I really did not flush out what the different between digital literacy and networked literacy were. I have shown this diagram in a few presentations and very few educators raised their hands when I asked if they were teaching digital literacy in their schools. I have also never seen a hand raised when I ask about network literacy.

Digital Literacy:

Wikipedia defines Digital Literacy as (http://en.wikipedia.org/wiki/Digital_literacy): ***Digital literacy** is the ability to locate, organize, understand, evaluate, and create information using digital technology. It involves a working knowledge of current high-technology, and an understanding of how it can be used. Digitally literate people can communicate and work more efficiently, especially with those who possess the same knowledge and skills.*

Microsoft defines Digital Literacy this way (http://www.microsoft.com/about/corporatecitizenship/citizenship/giving/programs/up/digitalliteracy/default.mspx): *The goal of Digital Literacy is to teach and assess basic computer concepts and skills so that people can use computer technology in everyday life to develop new social and economic opportunities for themselves, their families, and their communities.*

By these definitions digital literacy looks at understanding technologies and their uses. It's everything from understanding folder structures on a computer to being able to successfully use e-mail to communicate with others. Digital literacy focuses on the literacy needed to be literate with technology today. From copy & paste to understanding how to trouble shoot problems with an Internet connection, digital literacy is what it takes to operate a computer today.

Networked Literacy:

I couldn't find a definition anywhere on the web of what networked literacy is or looks like, but I think it's a literacy that we in the blogosphere talk about a lot. Networked literacy is what the web is about. It's about understanding how people and communication networks work. It's the understanding of how to find information and how to be found. It's about how to read hyperlinked text articles, and understand the connections that are made when you become "friends" or "follow" someone on a network. It's the understanding of how to stay safe and how to use the networked knowledge that is the World Wide Web. Networked Literacy is about understanding connections.

Becoming part of a network takes time. Growing connections and finding your place within the community does not happen overnight, but it can happen rather quickly if you are an active participant in the community for others.

Using a personal learning network within your workday means reevaluating your working habits. Many of the networking tools, such as Twitter and Facebook can easily be kept in the background, running and working and there when you have time and if and when you need your network. Becoming an active member of the Twitter community is as easy as leaving a browser window open and then refreshing the page when you have time to see what's new. In this way, your network waits for you to have time to catch up. Find something you like, read it, retweet it, or answer someone's question and go back to your work. The trick is using your personal learning network in those 5 or 10 minute blocks of time you get in your working day. Whether it be switching between activities or sending your students off to PE, these are those short moments of time when your PLN can be working for you and you can work to help others.

Some people describe their personal learning networks as virtual staff rooms. Think about the conversations that you have both personally and professionally in the staff room with colleagues, now imagine those conversations in the background of your classroom always available when you want them. Always working, only in this case you are not only talking to the colleagues in your school, but to colleagues around the world who have become part of your learning network.

Growing Your Professional Network

Growing your network to the size you want is not very difficult, it is the self-promotion part of the job that most people have an issue with. In order to grow your network you need to be recognized. Whether through your blog, a wiki project, Twitter, or a host of other social-networks, there is a sense of self-promotion you need to do. In this way you are able to grow your network and create the connections you want and that will work for you. It is much like opening the door to your classroom and having an "anyone's welcome" policy. Only now you are opening the door to your professional life online.

Many people have an issue with this and rightfully so. If you are going to be an active member of any community, there is a sense of self-promotion that needs to happen. I am not talking the "I'm the greatest" type of self-promotion, but rather the "This is who I am, this is what I do, this is what I'm interested in, and this is what I'm trying to learn" type of self-promotion.

It is like you are applying for a job, but rather in this case you are applying to people's interest. If your interests are the same or you feel you can help each other out, then a connection is or should be created. People need to understand who you are, what you do, and what you are about before they will create a connection with you, before they will count you as a node in their own learning network. For example, when you go to Wefollow.com to start building your Twitter network, you will read a couple of sentences that people have put on their profile. This is what they feel best describes them and why they use Twitter. If after reading their short blurb (a.k.a. bio), you think you might be able to learn from this person or think that this person is interesting, you reach out and create a connection to them and they become part of your learning network. That blurb or bio is an avenue for you (or others) to 'promote' yourself to the community (Figure 2.2).

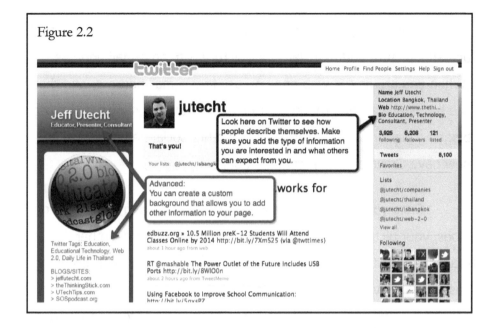

Figure 2.2

Part of growing your network and adding nodes, or people you can learn from, is learning how to brand yourself. By branding yourself (whether it's an image, a saying, or a website), it is important that people know who you are and recognize you when they see you. One place to get started is to create an avatar. An avatar is a small icon that represents you (Figure 2.3). It can be a picture of you, or a picture that represents you. I am sure you have see the little square images on websites before, pictures that people use to, you got it, promote themselves. When you leave a comment on a blog, sign up for Twitter, or have a profile picture on Facebook, you are creating a brand around who you are. Your avatar image is an important way that people will identify you.

Gravatar.com is "A Globally Recognized Avatar" and a great site to help you get started by uploading a picture that will become your globally recognized avatar for free. Gravatar.com allows you to upload a picture and associate that picture with an e-mail address. (When you are on the web and you use that e-mail address, and that site has Gravatar integration your picture, or avatar, shows up.) Most blogging platforms, wiki platforms and other social-networking sites now have Gravatar.com recognition built in.

Once you have a Gravatar and you start leaving comments on blogs, your picture starts to appear with your comments, other readers and authors start to recognize your avatar and your branding has begun. Having a common picture that you use on all social-networks is important so that you start to build an association with the image and your name. Much like Nike or McDonalds, your brand helps people recognize who you are. Take some time to think about the type of avatar you want to use to represent who you are on the Internet. You'll have to decide what type of picture represents you. Using a photo of yourself makes you recognizable to others if and when you meat people face to face. Some people choose to create a cartoon avatar of themselves (http://mashable.com/2007/09/12/avatars/), while others choose to just have a picture they feel represents them. Although you should take some time to consider what message your avatar sends to others, don't allow yourself to get bogged down in the details. You can always change it, and in fact most people change their avatar a few times when getting started until they find the picture that they feel best represents them.

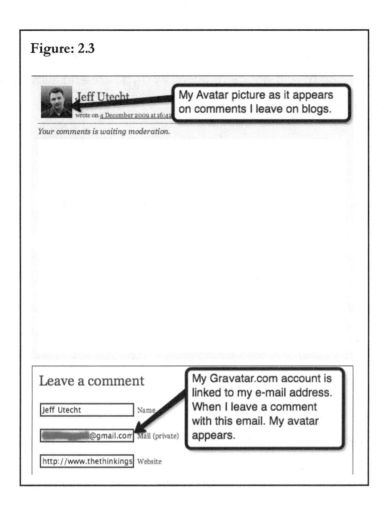

Figure: 2.3

Once you have your avatar it is time to start spreading your message. How do you do that? It's as simple as: Links, Links, Links!

In Figure 2.3 you can see in this comment I left on a blog that I added my blog address. That's how I create my network. Whenever I have a chance to type in the address to my blog, I do. Of course it does not have to be an address to a blog. It could be an address to a Twitter account, a Ning account, or a Facebook profile. The important part is to give people the opportunity to connect with you. Give them some place to go to find out more about who you are and allow them to start creating those connections.

Making your Network Work for You

When do you officially have a network? There is no magic number. A few people can be a network, or a few thousand. What makes it a network is when you start using the collective intelligence of others to find information, resources, and collaborate on projects. The interaction between you and the people you have connected with, or who have connected with you, is what creates a network. Once those connections are in place, you can start using your network to learn, hence creating a Personal Learning Network.

As the school day ended, Mr. Harter, a high school math teacher, headed to his desk to start working on tomorrow's lessons for his Algebra2 class. He has been trying to find ways to help some of his struggling students who are having a hard time with the new concepts he has been teaching. If only there was a web site that students could use that not only helped them to solve problems but also helped to explain the steps and the reasoning behind the solution. Although interested in the right answer, Mr. Harter tends to focus most of his attention on the process of solving the problem, not the solution itself. Understanding how and why you get a solution and being able to explain the procedure, he believes are more important.

After searching for 10 minutes for a web site that might help him, Mr. Harter turns to his PLN. He heads to his Twitter account where he has been slowly building a network of other high school math teachers. He simply types "Anyone know of a website that will help students solve math problems and explain the process of how the problem was solved?"

He then returns to working on his lesson plans, leaving Twitter open in the background. Mr. Harter has only been using Twitter for a couple of months after hearing about it at a recent conference. Although he's one of 6 high school math teachers in his school, Mr. Harter wasn't sure if there would be other math teachers on Twitter but he followed the directions and went to wefollow.com and typed in math teacher in the search box. He was amazed when he found there were 21 other math teachers out there sharing their knowledge that he could connect too. He took 30 minutes to read their short bios, visit their Twitter accounts to see what they were talking about, and then followed some that looked interesting. He was surprised when he started receiving e-mails from Twitter saying that some of the math teachers he had followed had followed him back. Now with a network of 15 or so math teachers Mr. Harter is finding it useful to connect with and share ideas with others around the world.

15 minutes after he submitted his question Mr. Harter gets the following reply:

"@dharter are you looking for something like this: http://www.wolframalpha.com"

Mr. Harter checks out the site and is instantly excited! This is exactly what he was looking for. He quickly heads back to his Twitter account and writes

"@mrmath Thanks! This is a fantastic site!"

Using your network to help you in your job is just the first step and where most people start, but it can also easily be used to help students learn and create connections for them around learning outcomes.

In Ms. Hellyer's 3rd Grade class they're getting ready to start their social studies lesson for the day. The class has been working on mapping skills in Social Studies and they are about half way through their weather unit in Science. Today's lesson should be fun if Ms. Hellyer's PLN is as active as normal.

Ms. Hellyer starts the lesson by handing out a blank map of the world to the students that just has the outlines of countries on it. First Ms. Hellyer does a mini-lesson focusing on prior knowledge and has the students create a compass rose in the upper left hand corner of the map.

As the students create their compass roses on the map, Ms. Hellyer turns on the projector that is connected to her computer and calls up a Google Map (maps.google.com). She then quickly heads to Twitter and sends out a short message to her network of educators around the world.

"Please twitter back the high temperature for today in Celsius with the name of your city. We're working on weather and mapping skills in class today. Thanks!"

Ms. Hellyer then lets Twitter do the work and spread the message. She has about 600 followers and she knows they come from around the world. She is not sure who will respond or where they will be from. She is hoping that she will get enough responses to help the students start asking questions of why the temperature varies in different parts of the world.

As she waits for the responses, she leads the class in a discussion around the essential question "What causes weather?"

After a quick 10 minute discussion, Ms. Hellyer explains the task:

"Today we're going to look at high temperature reading from around the world. I've called on some of my friends who live around the world to help us out. We'll read the tweets and then on your map you'll write in the temperatures for the day. I'll place the temperature on the map here on the board and you will have to look where that city is. Then using your mapping skills, see if you can locate where that city is on your blank map and write in the city name and the temperature."

Ms. Hellyer's PLN springs into action and by the time she returns to Twitter, she already has 5 people who have responded to her request.

@nzchrissy 22c today in Melbourne, Australia

@nzchrissy a chilly -3c today in New York, USA

@nzchrissy a beautiful 30c today in Bangkok, Thailand

@nzchrissy snowing and -1c today in Chicago, USA

@nzchrissy windy and chilly 4c today in Paris, France

The students start plotting their maps and have a short conversation about what might cause those temperatures in different parts of the world. In the 35 minutes of the lesson, Ms. Hellyer and her class plot 22 cities with temperatures. They'll continue to do this over the next three days adding more cities and more temperatures as they continue to study maps and the weather.

Finding ways to make your network work for you is the fun part. As you continue to build your network, become active within it, and see how others in your network use it, you will continue to get ideas on how to use your own learning network for your own professional learning and connecting students in your classroom.

3 Tools to Build a Community

Throughout this book so far I have mentioned some of the different tools you can use to start building professional communities and networks. In this chapter we'll look at some of the tools you can use to start building a community.

Blogs

Blogs and blogging is probably the quickest and simplest way to start building a community of learners to learn from. The hardest part of blogging is getting a steady stream of readers and those that leave comments.

There are a number of free services on the web to get you started including blogger.com, edublogs.org (specifically for educators) or my personal favorite, wordpress.com. With a few simple clicks you can be ready to go, no web coding needed.

If you are a little more adventurous and want to have more control over the look and feel of your site and your content, web-hosting services can be found as low as $6USD per month. Once you decide on a domain name (mynewblog.com) you can install free blogging software with a few simple clicks from your new web host and be up and running in no time.

What many first time bloggers do, and I highly recommend, is to start with one of the free services, see if blogging is for you, and start to build your community of readers. Once you have done that, you can always move your blog and content over to your own web space at a later time. Your community of readers, if engaged, will follow you.

Setting up your blog is the easy part I'm afraid, finding time to blog, blogging regularly and regularly commenting on other blog posts is the difficult part. Building up a readership when blogging means reading other blogs and leaving comments on blog posts that speak to you. Those comments are read by other readers of that blog, and some will click through to your blog and have a read.

Most bloggers also like to find out who is reading and commenting on their posts and will often follow links back to your blog to find out more about you and to read your blog posts as well. The more you comment on other blogs and leave the address to your blog in comments, the more connections are created. The more your blog becomes noticeable, the more people will read it.

There's no real science to blogging other than to reflect on what you're learning, be true to thy self, and blog as often as possible.

From: www.thethinkingstick.com/do-you-give-yourself-permission-to-reflect

Published On: Sept. 28, 2008 edited for this book

Do You Give Yourself Permission to Reflect

I've been thinking about reflection lately and how we use it in our classrooms. I can remember being in elementary school and being asked to reflect in a journal. Reflection is a great process...a proven process of learning (http://www.infed.org/biblio/b-reflect.htm). We've been asking students to reflect for years in education so one simple question:

Do you give yourself permission to reflect during the workday?

and another question:

Do your administrators give you permission to reflect during the workday?

I say during the workday because I truly feel if we are

to become better educators we need reflection time built into what we do. Too often we end up like Jenny (http://jennylu.wordpress.com/2008/09/27/100-e-learning-professionals-to-follow-on-twitter/)

When you spend a considerable amount of time learning about how we transform learning with the use of new tools, you find yourself online a lot. Most of this effort happens outside of my working day which impacts sleep, family time and time spent with friends.

And that's not good!

Why is it that educators place a high value on the reflective process yet do not give themselves permission to do it during their own working hours? Every educator has prep time. We use that time in a multitude of ways, yet how many of us set time aside just once a week to take 30 minutes or so and reflect?

You don't have to blog or even write. Reflecting can be reading an educational journal, it is sitting and staring out the window, or it is writing down your thoughts.

Andy Torris, an administrator, finds time in the back of the car when he's going from one campus to another in his "Dispatch from the Road" posts. Andy uses his working day time to reflect and write about his thinking (http://www.sentimentsoncommonsense.com/?p=98).

Newcomer to the blogosphere, David Hamilton, has an excellent post on reflection (http://principallyyours.edublogs.org/2008/09/22/on-reflection/) and the act of reflecting:

"But lest we forget, reflection is hard work. Whether we are sorting out our emotions and discerning personal values and attitudes, or discovering the shaky underpinnings of contemporary truths, reflection takes work, and, I would suggest, it takes practice. As I prepared to write this blog, I was amazed at how difficult it is to keep focused on a single abstract topic for stretches of time over several days."

Yes, reflecting is hard work! It takes practice, but more than that it takes time. Do you give yourself permission to reflect?

At the Learning 2.008 conference we had seven unconference sessions where participants could choose to go find a corner and reflect. Yet I have had conversations with people who went to the conference who said:

"I just wish I would have had time to sit and play with everything I was learning."

You did! You just didn't give yourself permission to sit and let it soak in. Instead it was more important for you to go to a session. Don't blame the conference, we gave you the time...you just chose to use it in a different way.

Isn't that what we do with prep time during our working day? We make a choice on how we are going to spend that time. We make the choice to answer e-mails, grade those papers, or update our Facebook status.

During the Shanghai EduBlogger Con, I was talking with some newbie bloggers who asked the question:

"Where do you find the time to blog?"

My answers:

I schedule it into my workday. When I am hired and again the first week of school, I tell my administration that I will be blogging during working hours. Blogging for me is about learning and reflecting. Blogging is not just writing, it is the act of reading, thinking, reflecting and writing. As a technology person in a school helping teachers, I need that time to reflect and learn about what's happening, and I make a point to schedule that into my workday.

Question:

"So you close your door, make yourself unavailable and blog?"

Answer:

Yes! I, unlike a classroom teacher, do not have set aside prep time. So I create my own around my lunch hour. I give myself 30 minutes of reflection time every day and back that with a 45 minute lunch. I shut off my e-mail client, I shut my door (or would if I had one) and I reflect. If a teacher comes to my door and needs my help, of course I help them, but that rarely happens during the lunch hour.

I don't write a blog post every day. Some days I read my RSS reader, other days I listen to a podcast or watch a YouTube video. Some days I follow links and learn, and other days....I blog.

I give myself permission to reflect. As a learner, I need that time, I understand how important it is to reflect and my administrators understand that it is legitimate use of my prep time.

Make reflection part of your workday. If it is something you try and do outside of school, it won't happen. There is rarely a time when I'm not thinking about education and technology...but it's my passion and I love it! Some teachers have other interests and that's great! But give yourself time to reflect on your practice. Make it a habit to reflect and make it a part of your workday.

Give yourself permission to reflect.....it's OK

If you need permission from someone then you have it from me. Tell your administrator that I say you need to take time to reflect. If they have an issue with it...they can contact me!

Wikis

Wikis are another great community building site. Many educators find using a wiki easier than blogging. Although the two have different purposes and uses, they both can be used to build a community. Where a blog is used to build a community around a person's thoughts and ideas, wikis are used to build communities around content.

Because wikis can be edited by anyone (if open to the public, or by invited guests only if private) they make it easy for people to build communities around any given subject, curricular area, or project.

Take the U Tech Tips wiki I started in 2008 as a place for educators to share their free or low cost software they can use in their schools. I was looking for a site to find a list of software that other schools were installing on school computers. After searching for and not finding the list I was after, I decided that I could create a wiki and build a community of other educational technology people that were probably looking for the same content. I set up the free site at http://wiki.utechtips.com and then started asking other educators to contribute to the list. I added resources that my school used, others did the same. Today when you go to the site, you will find a pretty comprehensive list of software that is free or low cost that schools can use. People continue to add and update the site as software changes. The community there is not defined but instead is in flux as people come and go as they need the content housed on the site.

Wikis are known for creating communities around content. Whether used with students or with other educators, wikis are a quick and easy way to start building a community around content and a given subject.

Ning.com

Ning.com allows you to set up a complete social networking site with a few clicks of a button. Think of Nings as your own private Facebook complete with forums, groups, picture upload, video sharing, personal pages, and the ability to customize the site to your look and feel.

There are a host of educational Ning sites started already for you to reach out and connect with. After you create an account at Ning.com, simply search for a network you might want to join. Here are just a few of the educational Ning communities already alive and well:

http://www.classroom20.com/

With over 37,000 educators already, it's a safe place to get started in joining your first community.

http://arted20.ning.com/

Art teachers gather here to talk about digital art and art education.

http://digitalwritingworkshop.ning.com/

A Ning devoted to teaching digital writing in the writers workshop model.

http://mathematics24x7.ning.com/

Math teachers unit to share resources and ideas on teaching math K-12.

http://smartboardrevolution.ning.com/

Have a SmartBoard in your room? Join 4,000 other educators who share resources, notebook files and tips and tricks about using the SmartBoard in the classroom.

These are just a few of the educational communities that exist on Ning that can help you feel comfortable in a networked community. I encourage you to join a Ning, if not one of the ones listed above then find another one that peaks your interest. If you can't find one that exists around your content or idea, then create your own and start spreading the word and build your own community.

In the end there are many different sites and places to build online communities. Sometimes the most difficult choice in getting started is deciding what tool to use when building a community. I encourage you to join an established community, watch how people interact, see how conversations flow, and then get involved. Once you become more comfortable within the online community setting, you'll understand the tools' strengths and limitations and be able to make a sound choice when starting your own community on the web.

4 Network Building Tools

Twitter

"I don't get Twitter!" is a phrase I've heard often from people who have heard of Twitter but have not used it themselves. Twitter is an interesting social-networking tool. The original premise behind Twitter was to tell others what you are doing. What Twitter has become is a host of niche networks of users around any given topic including education.

Twitter is more than just updates of what people are doing at any given moment. It is links, information, news, and answers to questions, all in semi-real time.

There are two sides to Twitter. There are those people you follow and those people that follow you.

Twitter is a great site to use to explain how the connections within a social-network are formed and just how powerful they can be in helping you in your daily teaching life.

Let's look at a common use of Twitter for those who belong to that community and have formed learning networks.

In my Twitter community I have 5,300 followers. In other words 5,300 people who have Twitter accounts have decided that the content I create within the community is worth knowing about and they would like to be informed each time I post something to the community.

At the same time I follow roughly 4,000 people. These are 4,000 educators, technology people, friends and acquaintances that I feel I can learn from within this community. So within Twitter there are those that follow me (my followers) and those I follow (who I'm following). They are mutually exclusive, as I do not have to follow the same people that follow me. In fact I know I don't, and I also know that there are people I follow who don't know I follow them. Confused? Let's break it down this way:

Followers: People who believe you have something worthy to say and want to follow what you are doing in your classroom, school, or district

Following: People you feel will add value to your job, profession, or everyday life

The way Twitter works is that you can send out a 140 character message to your followers. So I might go to my Twitter account and write:

"Does anyone have a good video that help explains landforms to 3rd graders"

I simply type the above statement into Twitter and it instantly goes out to 5,300 people. Some of those people will be on Twitter the same time I am and will see my 'tweet'. At this point they have a couple of options. They could:

1. Ignore my tweet, I'll never know they saw it
2. Respond to my tweet with a link to a resource that might help
3. Retweet my request out to their network

The 3rd option is the most interesting of the three. A 'retweet' is when you take something someone else has said on Twitter and forward it to the people that follow you. This is where the power of Twitter and social-networks truly take hold. Say for a second you have 300 followers, that is 300 people that like to get your updates and feel you are an added value to their daily lives. You 'retweet' (denoted by an RT on Twitter) my request out to your followers that will look like this:

"RT @jutecht Does anyone have a good video that helps explains landforms to 3rd graders"

Now my network of 5,300 educators is now 5,600 educators. Let's say that 5 people decide to retweet my request for a good video, before you know it my request has been seen by thousands of educators around the world all looking or potentially looking for a good video that helps explain landforms to 3rd graders for me. When was the last time you had thousands of educators helping you to find the best resources on the web?

I have sent out similar requests to the one above and had links shared with me up to 10 days after I put out the request. In most cases the links that other educators share with me are of better quality then I could find in a couple hours searching Google. Better still is the fact that the people that have sent me the link have probably used the resource themselves and know it to be of good quality and appropriate for the grade level, something that Google can not do.

You do not need a network of thousands of followers to take advantage of this type of professional development. Most of the educators on Twitter are there, just like you, looking and learning and understand **the more you give, the more you will receive**. Educators as a whole are very giving and seldom will you have a request go unanswered after you have built up your network. In order to build your network however, you need to be an active node for others to want to connect too. If you are not active, others will not add you to their learning network as we learn by both giving and receiving and you must do both to be a true member of an online community. A Twitter network with just 50 followers can have great potential. Find 50 people you think you can learn from and be active. They, and their followers, will start following you back, and that's how you start to grow your personal network.

After you have set up your Twitter account, finding the right people to follow is the next step. I suggest using a site called WeFollow.com. WeFollow is a Twitter directory of users. As a user you can add yourself to three sections of the site using Tags. Other users can then come and browse those tags and find people to follow.

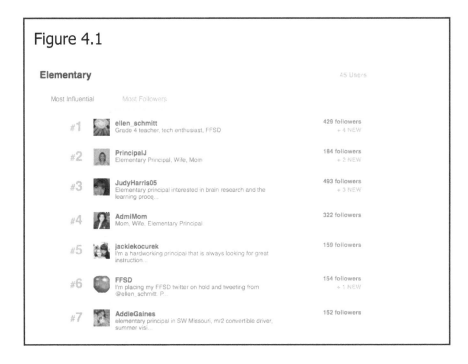

Figure 4.1

This is a screenshot (Figure 4.1) of the top 7 most influential elementary twitter users according to Wefollow.com a great place to start finding people to build your network.

The site http://wefollow.com/twitter/elementary has 33 users. If you are an elementary educator this might be a good place to start connecting with others. Or what about http://wefollow.com/twitter/internationalschools (one of my personal favorites) where you can find some 55 educators who work in international schools around the world. If you were looking to do a collaborative project and wanted to connect with another school, this might be a good place to look to make a connection. Of course it doesn't all have to be about education. A great way to get into using these tools is to connect with people that have a common personal interest as you. For example if you like jazz music try

http://wefollow.com/twitter/jazz

or how about

http://wefollow.com/twitter/dogs.

There are thousands of communities out there for you to build your network around. Joining the ones that will help you or give you pleasure both personally and professionally is a great way to get started.

When you follow someone on Twitter you are basically saying "Hey, I like what you have to say and would like to know what you are doing, thinking, and reading." So you go to their Twitter profile page and click the follow button. Soon after that anything that user writes on Twitter appears in your stream of tweets. You now get an update anytime that user posts anything on Twitter. Sounds a bit like RSS right? Well, it is fundamentally the same process. When you click

"Follow" on someone's Twitter profile, you create a connection to that person, so that anything he/she says appears on your Twitter page.

> Figure 4.2
>
> Love this image (via BlueSkunk) The Hierarchy of Digital Distractions http://is.gd/4U3JJ
> about 17 hours ago from Seesmic
>
> Just Blogged: Why Facebook is Unblocked at ISB http://bit.ly/2eMSmA
> about 18 hours ago from web
>
> Kids watching and commenting on each others teasure map movies. You can too! Http://inside.isb.ac.th/rm211 http://twitpic.com/pbi5q
> about 21 hours ago from Echofon
>
> One of our techies shared this with me today. An app for Mac that will tell you what apps have an update. http://is.gd/4T9wT
> 10:53 PM Nov 11th from Seesmic
>
> Fixing error on @thinkingchick blog which wasn't sending links of new blog posts to twitter.
> 3:16 AM Nov 11th from Seesmic
>
> A great night at the market $6 gets you enough food for three days. http://twitpic.com/p3gwc
> 2:34 AM Nov 11th from Echofon

If you were following me on Twitter, here are just a few of the updates you would have received from me (Figure 4.2). Like others on Twitter, I share links to interesting articles I'm reading and blog posts I've written. If you follow the right people they will do the web searching for you. All you have to do is click on the links. However, Twitter is a two way street, you must give to receive.

Of course people can do the same to your Twitter profile as well. They will search, or click on a link and land on your Twitter profile. If they like what you have tweeted recently, they will click follow and create a connection from your page to their page. Now they will start receiving any updates that you post on Twitter. In this way you create a network of people that you "follow" and they do the same. Sometimes you follow me and I follow you, other times I might follow you and you might not follow me. That's OK, I've decided I like your content; it doesn't necessarily mean you like mine.

After you have built up your network and you have 20 or so people following you, you can start using your Twitter network to find answers and information that you are looking for on the web.

Facebook

Facebook is the fastest growing social-network and a very powerful place to find resources and make connections with friends and family. Facebook connections are different than Twitter connections. Facebook relies on both parties being mutual "Friends". I can request to be your friend but until you say it's OK, I will not have access to your content in Facebook. Where the connection in Twitter is open to anyone, the connection in Facebook is closed and both parties must agree on allowing the connection to happen. Within Facebook, you know exactly who can see the content that you post because you have to give them access to see it. In Twitter you might not know who some of the people are that are getting your updates, but they have chosen to follow you for your content and there's not much you can do about it. This is the main difference between the words "Friends" and "Followers". In order to be friends you have to have a mutual agreement that you want to connect with each other and allow each other access to your content. Followers can follow you if they want, they might not know you, you might not know them, but the content that you create is your connection to each other. Some people feel more comfortable starting in Facebook and having the ability to control who can see their content. This sense of privacy that only your "Friends" can see your content is what has made Facebook so popular.

If you are already using Facebook (as I'm sure many of you are) to connect with friends and family, you might be hesitant to start building your professional network there. That's OK, it's your space you need to feel comfortable with it. One work around to this issue that many people are using is the use of Facebook Fan Pages. Facebook allows you to create pages that are more open to others within the Facebook community. By creating a Fan Page you can create a community around the content that you want to share openly within the Facebook community.

Many educators also struggle with the relationship with students on Facebook. As these relationship lines continue to blur, educators are going to have be more vigilant in understanding what content students have access to in communities such as Facebook. By creating a Fan Page in Facebook, educators can separate their personal Facebook lives with their professional ones. Great features of

Facebook Fan Pages are that no content overlaps between your Profile Page and your Fan Page. They are two completely separate sites (at the time of this writing).

Starting a Fan page is rather easy to do and it allows students to still feel connected to you, their favorite teacher, without you having to give up your privacy or be worried about what they might write or say that could end up on your wall. With a Fan Page, you can set the privacy settings so that fans can't post to your wall if you wish. These privacy settings allow educators to feel comfortable on the level of interaction they want to have with students and other Facebook community members. Fan pages also allow you to embed different widgets, not all the same widgets work on Fan Pages but you can embed your RSS feed and some other cool things to make your page...well....you. You can go to http://www.facebook.com/jeffutecht if you have a Facebook account to see how I have set up my Facebook Fan page to use with other educators and past and present students.

When students find me in Facebook and send an invite to be my "friend" I simply send them a message back that reads something like this:

> *I'm so glad you want to connect with me. I've decided to keep my Facebook profile private with my close friends and family. But I would like to stay connected to you. So if you would like to join my fan page here: http://www.facebook.com/jeffutecht we can stay in touch.*

Thanks!

...and then delete the invitation so that I know I've already invited them to my fan page.

I also think this helps to teach students that you don't have to become "friends" with everyone but that there are ways to stay private and it's OK to be private on the Internet. Again, the only way we are going to be able to teach students to be responsible on the Internet is for us, as educators, to set a good example and Facebook is a great place to start.

Using Facebook Groups

Jim Fitzgerald an IB English teacher at the International School of Bangkok was recently looking for a way to communicate with all of his 11th grade students. He taught three periods of IB 11th grade English and needed a way to update all the students instantly if a due date had changed, or if he had found further information about a topic they were discussing and wanted to share it with them.

He asked his class what system they would

recommend. Without hesitation the students said that a Facebook group would be the best tool.

With the students' help, Mr. Fitzgerald set up a Facebook group and then invited three students that he trusted and wanted to help to be administrators of the group. These administrators then invited all the other students to be members of the group. Within a day the group was created and all students had joined the group as a member. Now whenever Mr. Fitzgerald wants to communicate with all his 11th graders, he simply goes to Facebook and posts an update to the group, or he can message all of the students at one time or message them individually.

Much the same way teachers use e-mail and e-mail groups, Mr. Fitzgerald has found a way to use the community that the students were already a part of and created a network for his class within it. I asked the students why they liked using the Facebook group. One student replied:

> "We're always on Facebook anyway so it's great to know what is happening in class. It's the only way I knew that our assignment due date had been changed this week."
>
> What a great use of the tools already being used by students and finding a way to make it work in education.

Skype

As we continue to look at online communities and networks, one cannot over look Skype (Skype.com) the free international video telephone and chat service that has taken the world by storm. With more that 500 million users, Skype is one of the largest online community on the Internet today. Although many people do not think of it as a community, it's actually a very powerful one. Like all other community/network relationships we've looked at so far, Skype is one where the network really is where the power lies. None of us really care that we belong to a community of over 500 million, but what we do want is those contacts, that network, that allows us to telephone each other for free with video or via chat.

Skype is an amazing tool that many podcasts use to record conversations that they then turn into a podcast and release to the public. Because it's free and the quality continues to improve, Skype in 2009 made up 13% of all International calls making it the largest International call provider between countries. (http://www.itwire.com/telecommunications-news/international/30579-13-percent-of-international-calls-now-go-via-skype). That only accounts for people calling from Skype to another Skype account which is free. It does not take into consideration the Skype Out feature that allows users to call from Skype to a landline or mobile phone for pennies on the dollar of regular phone carriers.

What Skype continues to create is a way for people to connect at low or no cost. Skype is another great tool to add to your Personal Learning Network tool belt. The ability to connect with other professionals around the world, to connect classrooms across oceans, and to be able to video conference for free is simply amazing. From 2006 – 2008, I taught graduate courses for Plymouth State University while living in Bangkok, Thailand. The online courses were enhanced by my ability to either chat, call, or video students in the course when they had a question or just to check in on their progress in the class. There is nothing like your first International Skype call when it's time for bed in one place in the world and you are just waking up in another.

Getting started with Skype is simple. Just go to Skype.com, sign up for an account, download the Skype software, sign into your account via the software on your machine and start building your Skype network. This entire process will take you about 5 minutes to complete. Finding people on Skype is probably the most difficult part of building your network, as Skype is viewed by many as similar to a traditional telephone number that you don't just give out to anyone. Skype, unlike Twitter and Facebook which allow you to post to many people at once, still works like a phone and is a connection between two individuals. Because of this it's seen as more personal. I have a direct line to you, and you to me. Although you can have up to 10 people in a conference call on Skype, it's still that personal one-on-one telephone type system that makes it popular. Below is an example of how Skype as a network can work for you.

I was sitting at my desk working on a blog post when Skype popped up on my screen with a chat message from Chris Betcher (http://chrisbetcher.com) who lives in Australia.

The Skype Chat went like this:

>Chris Betcher: Jeff, how are you with Moodle? I have a Moodle problem that's driving me mad...

>Jeff Utecht: I'll give it a shot

>Jeff Utecht: What's the issue?

>Chris Betcher: Can you go to (URL) and tell me what you see?

>Jeff Utecht: Posted file Screen shot (I took a picture of my screen of what I saw and sent it to him via Skype file transfer)

Jeff Utecht: A nice looking login page :)

Chris Betcher: But just a login page right?

Jeff Utecht: Yep

Chris Betcher: No real content

Jeff Utecht: Nope

Chris Betcher: Are you able to do voice for a sec?

Jeff Utecht: Give me a sec to get it set up

Chris Betcher: K thanks

I then called Chris and after a 5 minute conversation we figured out the security setting that was not allowing content to show on the front page of his site. A 5 minute Skype call between Australia and Thailand fixed the issue.

Why would Chris use Skype instead of say Twitter or Facebook in this case?

Chris had a specific problem that he was working on at that time. Could he have gone to Twitter and posted his problem? Sure, but trying to explain what the issue was in a 140 character limit that Twitter has would not have been efficient information for someone to try and help him. How about Facebook? Sure, but you never know

69

who's on at that specific moment of time and his issue needed to be fixed sooner rather than later.

So Chris decided to use Skype. He opened up his Skype application, signed in and looked through his Skype list of other users who were on Skype at that exact time. He then narrowed down that list of people in his network that might be able to help him with his issue. He then sent me a chat, that led to a quick call, that then led to him fixing the issue he was having with his website.

This is the power of Skype. Skype's ability to show you what other users are on at that given time allows you to chat, call, or video with people when you need them, or when they need you. Skype also has other features that make it a powerful networking tool. The ability to transfer files in real time to a Skype contact, and the ability to share your screen with another user are two other important uses of Skype.

If we look at where Skype fits into your overall Personal Learning Network, it would be the most private of all the networking tools you use. Where Facebook is about mutual friends, and Twitter is about Followers. Skype is about personal connections. Where Facebook and Twitter are about you broadcasting to many, Skype is about one-on-one or small intimate group discussions and conversations.

> When using Skype it is good Skype etiquette to send someone a chat message first to see if they are available to talk rather than just calling them straight away.

Google Buzz

In February 2010, Google released Google Buzz, a social component to it's popular e-mailing platform Gmail. Google Buzz uses the contacts in your gmail account to automatically connect you to others using Buzz.

Google Buzz falls somewhere between Facebook and Twitter and at the time of publishing. It's still uncertain the impact Google Buzz will have on the social-networking space. Google Buzz allows you to post longer messages to followers, unlike Twitter's 140 character limit. It is also more open than Facebook's closed community. Anyone can search for your profile in Google Profiles (see mine here: http://www.google.com/profiles/jutecht) and then follow your Buzz updates. Much like Twitter, Buzz "Followers" and those that you "Follow" are mutually exclusive.

Similar to Facebook, Buzz allows you to comment on other people's updates, or even just give it a thumbs up that you like the update/information by clicking a button. It has the added feature because of its integration with gmail to e-mail any Buzz message to someone else.

The verdict is still out on whether Google Buzz will be a viable social-network that people migrate too. I mention it here in this book as something for you to look into and see for yourself if it has a place in your Personal Learning Network.

5 Connecting It All Together

If you have made it this far into the book, then you are probably feeling a bit overwhelmed and asking yourself "Do I really need all these ways to connect to people?" The answer is no, you don't need them all. First what you need to do is find a way to bring them all together, to understand that each network has a specific place, and performs a specific task, within your Personal Learning Network. Secondly, there are tools out there that will help you connect these social-networks together and help you manage your social-networks so that they do not feel overwhelming.

Living in a Cloud

Cloud computing is a fairly new concept that helps us to connect all these systems together. Cloud computing is the idea that everything we do is stored "in the cloud" or in other words not on your computer or a physical hard drive. The first cloud computing system you may have used was e-mail. Whether you had an AOL, Hotmail, Gmail, or Yahoo e-mail account, you were using server space by one of those companies to hold all your data. When you wanted to check your Yahoo account, you signed into that account and had access to your e-mails and any attached files you had stored there. In other words everything is stored on the Internet.

As the cost of servers and server storage continue to decrease, companies such as Google, Yahoo, Microsoft and Amazon have been able to store more of your data for you. What these companies, and others like them, have done is create large server farms, or large buildings full of servers, around the world to hold all of your data. Today, you can upload 2 Gigabytes (GB), or 3CDs worth of information, of video files to YouTube.com without thinking about it and allow Google to store that information for you. That's 2GB of space you now have free on your computer for other information. Or upload all your pictures to Flickr.com and use Yahoo's storage space to save all your pictures, again not having to take up storage space on your personal computer.

The added benefit to this is not only do you not have to keep buying hard drives to store your own data, but once that data is "in the cloud" or on a server somewhere in the world, it's accessible anywhere you have an internet connection.

This book was written using Google Docs (http://docs.google.com) a cloud based word processing program that allows me to work on the book from any computer that has an Internet connection. I just log into my account and start working. Because the book is in the cloud, I get the added feature of easily being able to share it with others. Once I give my editors access to the files in Google Docs, they will be able to view and edit the book from any Internet capable computer in the world.

Although cloud computing is mostly about storing data on the Internet, what has come of this is the ability to easily share files with others quickly and without having to worry about space. This is what has helped to make Facebook so popular. The ease of which people can upload pictures and videos to Facebook makes it a great place to save those files, but what really makes it powerful is the ability to then share those pictures and videos with the people you want. No more having to e-mail or have multiple copies of the same picture flying around the web. Upload all your vacation pictures to one spot and just invite people to see them.

Because of these connections that can be made when things are kept on the Internet, it is easy for users to start connecting social-networks together, or use applications on your computer that will connect them together for you.

If you want you can connect your Facebook updates with your Twitter updates. So each time you post something on Twitter it automatically posts it to Facebook for you at the same time. You post once and the connections in the cloud do the rest for you. Or using Google Buzz you can import your Twitter updates, your YouTube videos, your Flickr pictures, and more into one stream. So an update to any of those sites will automatically be shared with your Google Buzz followers if you allow it to.

Understanding the connections that can be made when all of your content is web based, is a powerful idea worth taking the time to wrap your head around. Once the content is on the web, it can be pushed and pulled to other places on the web. Each time I write a blog post that content gets pushed out to both Twitter and Facebook. My Twitter account is connected to my Google Buzz account so Google Buzz pulls the content from Twitter into its social-network.

From: www.thethinkingstick.com/get-twitter-off-the-web-and-on-your-desktop

Published On: June 1, 2009 edited for this book

Twitter is a playground to me. It has been since I started using it in 2007 (man that seems so long ago!). It's been fun to watch it grow over the past couple of years and hit the mainstream in the past couple of months. I'm excited to see where it goes, but in the mean time let's keep playing!

I downloaded the Seesmic Desktop Application (http://desktop.seesmic.com/) a couple weeks ago and had a play with it. I enjoyed my time
with TweetDeck (http://www.tweetdeck.com/)and still haven't uninstalled it, but after Kim had me try Nambu which is a Mac specific Twitter Desktop App, it sent me on a search for a new one.

Many people don't get Twitter. Even after signing up for an account people still have a hard time understanding how it works, or why you'd want to use it. I think using a desktop app is the only way to truly understand Twitter and use it successfully.

Because Twitter has gone mainstream, it makes it a nice news feed, friend feed, and information silo. Using a desktop app like Seesmic allows you to create groups of users, follow search terms, and keep track of a lot of different content easily. Seesmic has a great set of videos to get you started.

For example I have a column that follows any mention of the Mariners (it is baseball season after all!). I have a couple different columns of users. People often ask me how I keep up with so many followers. My answer is simple….I don't. I quit trying to keep up with them a long time ago and instead I've taken a data mining approach to Twitter and I've created groups that work for me.

I have a group called "My PLN" if you are in education and you have shared a link that I've enjoyed or happened across you get put into this group (2 clicks and 2 seconds). I have a group called "My Peeps" these are my close friends and colleagues I work with. It allows me to keep a close eye on say Dennis Harter (http://www.dennisharter.com) my colleague and what he's tweeting about.

The other nice feature of Seesmic and most Twitter Desktop Apps is that it allows me to easily unfollow or block spammers, making it easy to manage my profile from within the Twitter App itself.

Then there is the added bonus of being able to update your Facebook status within the same application, now if only they could add FriendFeed updates...they'd have the trifecta!

On my Mac I run Seesmic in a Space. It gets a space all to itself and simply runs in the background. I go to it when I need something, have a second to catch up on a conversation or need information on something happening in real time. I go there when I have something to share, when I feel like I can add to the value of other's stream of information or when I simply want people to know what I'm doing....after all....that was its original use.

TwitterCamp still has the most potential

TwitterCamp (http://www.danieldura.com/downloads/TwitterCamp.air) was the first AIR app (http://www.adobe.com/products/air/) that I installed and today I still think it has the most potential for use in schools and conferences.

Cloud Computing = Mobile Devices

It is not a coincidence that mobile computing (i.e. cell phones, netbooks, iPad) has taken off in recent years as cloud computing has become more popular. Once everything is stored on the Internet, accessing it from anywhere, with any type of Internet enabled device becomes easy. Cloud computing has lead to the success of the iPhone and other mobile devices including e-readers such as the Kindle and the Apple iPad. An iPhone developer can now create a program and with a touch of a button access information on the web. If all the information you need is stored on the web, then creating cheap devices to access that data stored there is not that difficult. The Amazon Kindle works on the same principle. You can order a book, or receive the New York Times from the Internet and download it to your device. You don't have to be connected to the Internet all the time, but when you want your paper, or you need to buy a new book to read, it's there waiting for you.

Cloud computing has also brought a newer cheaper computer to the market. Coined "Netbooks" these small inexpensive machines are made to do one thing and do it well-connect to the Internet. They have very little storage room on the actual computer and are made to do nothing more than connect you to the Internet where you can get work done. Because online social-networks are all web-based, using mobile devices to help manage the content there is easy. Simply download the Twitter and Facebook application to your iPhone and with a touch of a button, no matter where you are, you are connected to your personal learning network. Of course you don't have to be, but it is there if you need it.

Finding ways to connect your social-networks together won't only save you time, but will also allow you to see connections between people and content. Find a system that works for you. Personal Learning Networks are just that, personal. What works for me might not work for you. Spend some time and find a way to connect your networks together to create one large network that talks to each other the way you want it to on the devices you use the most.

6 Making It Personal

When it is all said and done, creating your PLN starts with you. You have to make it personal, and you have to make a commitment to giving it a go. Much like good eating habits or a regular work out schedule, it takes time and it usually means altering what you've done in the past. Creating, maintaining, and using your PLN effectively will take time and effort, but in the end I'm convinced that you'll find value in it personally and professionally.

I have noticed an emerging trend of what one goes through when starting a PLN for the first time. I continue to look at the stages I am going through in adopting this new way of learning, interacting, and teaching in a collaborative, connected world.

As I have helped others start their PLNs I have found that many of them go through these same stages. Stages I have coined as the "Stages of Personal Learning Network Adoption."

Stages of Personal Learning Networks Adoption:

Stage 1 Immersion: Immerse yourself into networks. Create any and all networks you can find where there are people and ideas to connect to. Collaboration and connections start to take off.

Stage 2 Evaluation: Evaluate your networks and start to evaluating which networks you really want to focus your time and energy on. You begin feeling a sense of urgency and try to figure out a way to "know it all."

Stage 3 Know it all: Find that you are spending many hours trying to learn everything you can. Realize there is much you do not know and feel like you can't disconnect. This usually comes with spending every waking minute trying to be connected to the point that you give up sleep and contact with others around you to be connected to your networks of knowledge.

Stage 4 Perspective: Start to put your life into perspective. Usually comes when you are forced to leave the network for a while and spend time with family and friends who are not connected (a vacation to a hotel that does not offer a wireless connection, or visiting friends or family who do not have an Internet connection can facilitate this!).

Stage 5 Balance: Try and find that balance between learning and living. Understanding that you cannot know it all, and begin to understand that you can rely on your network to learn and store knowledge for you. A sense of calm begins to settle in as you understand that you can learn when you need to learn and you do not need to know it all right now.

Of course balance is something we are all trying to achieve in our lives, and finding where your PLN belongs in a balanced life is something for constant reflection.

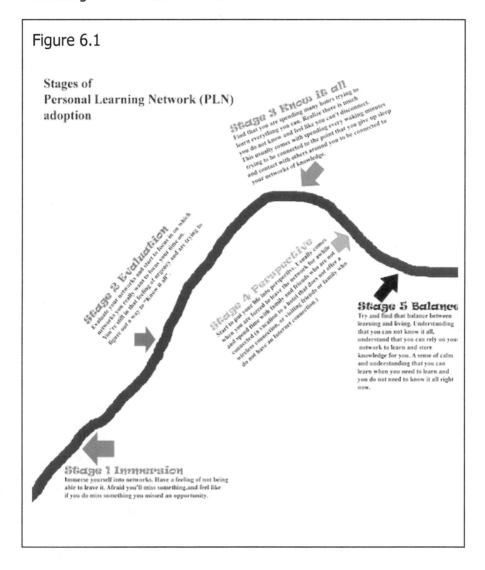

Figure 6.1 shows a correlation with learning. As you immerse yourself into the network, your learning increases. The more you learn, the more you want to learn and the more immersed you

become within the network. You then reach a point where you feel you can stay on top of the information you are interested in. Instead of feeling "behind" you feel as though you can keep up. You understand how the tools work and use them in your own personal learning.

I do not believe you have to go through all these stages. Some people jump from stage 2 to stage 5 or do not become so immersed into their PLN that they ever reach stage 3, that sense of having to "know it all." There is no scientific data behind these stages, just what many have seen and felt as they have started PLNs of their own. If nothing else, I hope this helps to frame some thoughts in your head around your own PLN journey.

In the end you need to remember it is about you. You choose the communities to join. You choose who is in your network; you build your network to work for you. If you find that you are not getting anything out of your network, then it's time to reevaluate whom you are connected to and if they are the right people for you. People sometimes feel bad when they "unfriend" or "unfollow" someone. Just remember it's your network not theirs. You need to be connected to the people you want to connect to, that you want to communicate with, keep in touch with, and learn with.

Look for communities and networks not only online but in the real world as well, and think about the relationships between them. Creating and building a Personal Learning Network might just be the best professional development you have ever had and it is all out there waiting to be connected to you.

How Far is your Reach?

Building learning communities and networks online means reaching beyond the walls of your classroom, the walls of your school, and even the walls of your own state, country, or continent to create connections. Once you reach out and make those connections with other educators, professional learning like you have never experienced before starts to flow your way. You will soon find yourself in a state of continual learning. Taking advantage of the constant stream of information available today on the web at any given moment and using it for your own professional growth is what it means to learn in the 21st Century. It is the ability to connect to the information and people you want to learn from. It is literally having the resource that is the Internet at your fingertips when you need it.

It can only happen if you reach out and start making connections. There are other educators out there waiting to connect to you and wanting to learn from you. If you reach out to them, some of them will reach back. New learning networks are formed, new relationships are fostered and before you know it, learning in a networked world is just part of who you are. Now is the time to get started, join a community, create a network and start reaching out to others.

References and Links

Ahonen, T. (2008, December 10). Trillion with a t, the newest giant industry has arrived: the money and meaning of mobile. Retrieved from http://www.communities-dominate.blogs.com/brands/2008/12/trillion-with-a.html

Beckert, S. (2010, January 19). International phone traffic growth slows, while skype accelerates. Retrieved from http://www.telegeography.com/cu/article.php?article_id=31718&email=html

Mayo, K, & Newcomb, P. (2008, July). How the Web was won. Retrieved from http://www.vanityfair.com/culture/features/2008/07/internet200807

Niman, H. (2009). Tracking the progress of h1n1 swine flu. Retrieved from http://flutracker.rhizalabs.com/

Wikipedia, . (2010, April 21). Prosumer. Retrieved from http://en.wikipedia.org/wiki/Prosumer

Avatars
An avatar is a small logo or picture that represents who you are on the Internet.

Gravatar.com
Globally Recognized Avatar: A website where you can associate an e-mail address with an avatar. Anytime you use that e-mail on the Internet the site can pull your avatar picture from Gravatar.com and display it on their web page. (Page 33, 34)

http://mashable.com/2007/09/12/avatars
A list of sites that can help you create your avatar. (Page 34)

Blogs
A website that allows readers to comment and communicate with the author.

http://www.sentimentsoncommonsense.com/?p=98
A blog post by Andrew Torris titled: <u>Dispatch from the Couch: Reflecting on Reflecting</u> published September 27, 2008. (Page 45)

http://principallyyours.edublogs.org/2008/09/22/on-reflection/
A blog post by David Hamilton titled: <u>On Reflection</u> published September 22, 2008. (Page 45)

http://jennylu.wordpress.com/2008/09/27/100-e-learning-professionals-to-follow-on-twitter/
A blog post by Jenny Lu titled: <u>100+ (E-) Learning Professionals to follow on Twitter</u> published September 27, 2008. (Page 44)

http://www.dennisharter.com/
Dennis Harter is the Dean of Students at the International School Bangkok. (Page 77)

http://www.infed.org/biblio/b-reflect.htm
Mark K. Smith's short essay titled: <u>reflection</u> published in 1999. (Page 43)

http://chrisbetcher.com
Chris Betcher is a teacher, he also supports teachers in integrating ITC skills in their classroom. (Page 68)

Computation Search Engine
Web-based search engines that not only search for content on the Internet but can also solve sophisticated mathematical algorithms

http://www.wolframalpha.com
A website that can figure out sophisticated mathematical algorithms as a well as many other functions that deal with numbers. It is still in beta but is making waves in the world of search engines. (Page 37)

Delicious
A social bookmarking website that allows you to save your bookmarks or favorite websites so that they are accessible anywhere you have an Internet connection.

http://www.delicious.com/tag/globalwarming
A search for the tag globalwarming on delicious. (Page 20,21)

Digital Literacy
The term given to the idea that paper-based reading is different than computer or digital based reading. If paper-based reading is literacy than digital literacy refers to the reading and comprehension of content in a digital format.

http://www.thethinkingstick.com/digital-literacy-vs-networked-literacy
A blog post by Jeff Utecht titled: Digital Literacy vs Networked Literacy published on August 7, 2009. (Page 27)

http://www.ncte.org/press/21stcentwriting
A report by Kathleen Blake Yancy from the National Council of Teachers of English titled: Writing in the 21st Century. (Page 27)

http://www.ncte.org
The National Council of Teachers of English website. (Page 27)

http://www.thethinkingstick.com/the-age-of-composition
A blog post by Jeff Utecht titled: The Age of Composition published on March 16, 2009. (Page 28)

http://www.microsoft.com/about/corporatecitizenship/citizenship/giving/programs/up/digitalliteracy/default.mspx
Microsoft Corporation's definition of digital literacy. (Page 29)

Facebook
The most popular social-networking site in the United States as of this writing.

http://www.facebook.com/jeffutecht
Jeff Utecht's Facebook page where students and those looking to connect with Jeff Utecht can join the page. (Page 63)

Google
The most popular search engine in the United States.

http://www.google.com/profiles/jutecht
Jeff Utecht's Google Profile Page. (Page 71)

http://docs.google.com
Google Docs is an online office suite of tools (word processing, spreadsheet, presentation and storage) free for use by everyone. (Page 74)

Ning
Ning.com is a website where you can create your own social-network or join the social-networks created by others.

http://www.classroom20.com/
The largest of the educational ning social networks. (Page 51)

http://arted20.ning.com/
A social network for art educators at all levels. (Page 51)

http://digitalwritingworkshop.ning.com/
A social network for educators interested in digital writing and the writing workshop model. (Page 52)

http://mathematics24x7.ning.com/
A social network for math educators. (Page 52)

http://smartboardrevolution.ning.com/
A social network for educators with SmartBoards or who would like to learn more about using them in the classroom. (Page 52)

RSS

http://www.google.com/reader
Google Reader is a free RSS reader offered by Google. You will need a Google account to set up and access the site. (Pages 12,13, 22)

- *http://tinyurl.com/googlereaderintro*
  Google Reader Intro video

- *http://tinyurl.com/googlereadersetup*
  A video about getting started with your Google Reader

http://www.pagefllakes.com
A free widgetized start page that allows you to customize the widgets you want displayed on your personal home page. You can add RSS feeds from your favorite websites along with a large variety of built in widgets. (Page 11,13)

- *http://tinyurl.com/pageflakesintro*
  An introduction video to Page Flakes
- *http://tinyurl.com/pageflakesshare*
  A video to help you get started and show you how to share your Pageflakes with others.

http://www.netvibes.com
A free widgetized start page that allows you to customize the widgets you want displayed on your personal home page. You can add RSS feeds from your favorite website along with a large variety of built in widgets. (Page 11,14,15,24)

- http://tinyurl.com/netvibesintro
  An introduction video to Netvibes
- http://tinyurl.com/netvibessetup
  A video to help you get started with Netvibes

Skype
A Voice Over Internet Protocol service that allows you to call other Skype users for free from computer to computer and very inexpensively from computer to landline or cell phone. It also includes video calls and group video calls (VOIP).

- http://www.itwire.com/telecommunications-news/international/30579-13-percent-of-international-calls-now-go-via-skype
  An article from itwire.com titled: <u>13 Percent of International Calls now go via Skype</u> published on January 20, 2010. (Page 67)

Twitter
A social-networking site that allows you to send out updates and requests in 140 character chunks.

http://www.wefollow.com
A website that allows a Twitter user to add themselves to a category and allows other Twitter users to find people to follow in categories they are interested in. (Page 32, 27, 58)

- *http://wefollow.com/twitter/elementary*
  A category of Elementary Educators using Twitter.
- *http://wefollow.com/twitter/internationalschools*
  A category of International School Educators using Twitter.
- *http://wefollow.com/twitter/jazz*
  A category of Twitter users who like to talk about Jazz music.
- *http://wefollow.com/twitter/dogs*
  A category of Twitter users who like talking about dogs.

http://www.thethinkingstick.com/get-twitter-off-the-web-and-on-your-desktop
A blog post by Jeff Utecht tilted: <u>Get Twitter off the Web and on Your Desktop</u> on June 1, 2009. (Page 76)

http://desktop.seesmic.com
Seesmic is a Twitter desktop application that allows you to view and tweet from your desktop instead of via a web browser. (Page 76)

http://www.tweetdeck.com
A desktop application that allows you to keep track of your Twitter followers and create your own groups. (Page 76)

http://www.danieldura.com/downloads/TwitterCamp.air
Twitter Camp is an application that shows a set of tweets based on your criteria in a very visually stimulating way. It's a great application for conference or large venues. (Page 78)

http://www.adobe.com/products/air
A program created by Adobe that runs many desktop applications that need to connect to the Internet to get information. Many of the most popular Twitter applications run on Adobe Air coding. (Page 78)

Wikis
Wikis are websites that are easily editable and allow users to share their content and editing rights with others.

http://wiki.utechtips.com
A wiki created by educators around the computer programs they use with students in schools. (Page 50)

Wikipedia

http://en.wikipedia.org/wiki/Web_2.0
A working definition of the term Web 2.0 (Page 1)

http://en.wikipedia.org/wiki/Digital_literacy
A working definition of the term Digital Literacy (Page 28)

Reach Website

Information in this book will change due in part to the changing nature of technology and the tools available to use today. I have set up a website to help guide you further on this journey in online communities and networks. Here you will find "How To" videos, downloadable PDFs, and links to other resources on the web to help get you started. You can also leave a comment or a question about the book and join the community of other readers of Reach.

www.jeffutecht.com/reach

About the Author

Jeff Utecht is an international educator and an educational technology consultant. He began his career in Elementary Education and subsequently earned his Master's Degree in Curriculum and Instruction with a focus on technology. Originally a fourth grade teacher from Washington State, Jeff was awarded a Bill and Melinda Gates Technology Leadership Grant in 2000 that put 7 computers and a projector in his room. This was the start of his passion for using technology in education. Now 11 years later he speaks and consults for schools and organizations around the world.

Over the past 8 years, Jeff has lived and worked in Saudi Arabia, China and currently resides in Thailand. His global experience in education has allowed him to see first hand how the Internet is changing education. An early adopter with blogging, his blog thethinkingstick.com has been featured in a number of books and articles as well as being used as a continual resource for educators around the world. Jeff has taught online graduate classes since 2007 for different American Universities and continues to teach courses online in the area of educational technology.

Jeff is currently a Technology & Learning Coordinator for the International School Bangkok, where he works with both educators and students in understanding the social web we now live in.

Made in United States
Troutdale, OR
06/17/2024

20598178R00060